Classic Deere Tractors

An Album of Favorite Big Green Farm Tractors from 1914–1970

By Cletus Hohman

Voyageur Press

Copyright © 2003 by Voyageur Press, Inc.

All rights reserved. No part of this work may be reproduced or used in any form by any means—graphic, electronic, or mechanical, including photocopying, recording, taping, or any information storage and retrieval system—without written permission of the publisher.

Edited by Michael Dregni
Designed by JoDee Mittlestadt
Printed in Hong Kong

03 04 05 06 07   5 4 3 2 1

Library of Congress Cataloging-in-Publication Data

Hohman, Cletus.
  Classic Deere tractors : an album of favorite big green farm tractors, 1914–1970 / by Cletus Hohman.
     p. cm. — (Farming legends)
  Includes bibliographical references and index.
   ISBN 0-89658-620-0 (hardcover)
  1. John Deere tractors—History. 2. Farm tractors—United States—History. I. Title. II. Series.
  TL233.6.J64 H64 2003
  629.225'2'0973—dc21
                        2002151904

Distributed in Canada by Raincoast Books, 9050 Shaughnessy Street, Vancouver, B.C. V6P 6E5

Published by Voyageur Press, Inc.
123 North Second Street, P.O. Box 338, Stillwater, MN 55082 U.S.A.
651-430-2210, fax 651-430-2211
books@voyageurpress.com
www.voyageurpress.com

Educators, fundraisers, premium and gift buyers, publicists, and marketing managers: Looking for creative products and new sales ideas? Voyageur Press books are available at special discounts when purchased in quantities, and special editions can be created to your specifications. For details contact the marketing department at 800-888-9653.

Legal Notice
This is not an official publication of Deere & Company. The name John Deere and Lanz, as well as certain names, model designations, and logo designs, are the property of Deere & Company Inc. We use them for identification purposes only. Neither the authors, photographers, publisher, nor this book are in any way affiliated with Deere & Company.

On the frontispiece:
Ma waves from the kitchen to Pa aboard his Poppin' Johnny as Junior waves his rattle.

On the title pages:
John Deere Model BNH. Owners: Walter and Bruce Keller. (Photograph © Andy Kraushaar)

Inset on the title pages:
A youngster rides his Deere pedal tractor with pride.

Inset on the contents page:
Pa and Rover head for the fields with the family's Johnny Popper.

Contents

Introduction
Farewell Horses, Hello Tractors   7

Chapter 1
Deere Enters the Tractor Field, 1914–1923   9

Chapter 2
Poppin' Johnnies, 1923–1937   21

Chapter 3
Tractor Style Down on the Farm, 1938–1955   43

Chapter 4
Lanz, the German Cousin, 1911–1970   61

Chapter 5
The Numbered Series Years, 1952–1960   69

Chapter 6
Big Green and the New Generations, 1959–1970   83

Bibliography   92

John Deere Clubs, Magazines, and Resources   93

Index   95

Above: **"Vittles for Victory"**
A Deere graces the cover of a World War II–era issue of *American Power Farmer* magazine.

Facing page: **John Deere Model GP**
Launched in 1930, the third version of the Model GP was known as the "Crossover" as its manifold and air intake ran through the hood. Starting with the Series 3 and continuing to the final Series 5, these later GP tractors featured a larger, 339-ci (5,553-cc) engine with 25 hp instead of the earlier 312-ci (5,111-cc) powerplant with its 20 hp. Owners: Derrick and Darold Sindt. (Photograph © Andy Kraushaar)

INTRODUCTION

Farewell Horses, Hello Tractors

The Arrival of a Revolution in Farming

From the dawn of agriculture to the mid 1800s, advances in farming came as slowly as a horse or oxen team pulling a plow through the prairie soil. Radical new advances in reaping and threshing were reined back by the lack of power to realize their full potential. Farmers still plodded along at a horse's pace.

The arrival of mechanical horsepower was the key to the future of farming. With the development of steam engines, the Industrial Age arrived on the farm in the 1880s. Steam engines could power separators, and with the development of drivetrains, steam traction engines pulled plows.

The world of farming was forever changed at the dawn of the 1900s when kerosene- and gasoline-fueled internal-combustion engines replaced the steamers. Several decades of engineering and field work hammered the unreliable and eccentric early tractors into shape, and by the 1920s, gas tractors were slowly finding a place on most every farm in the nation.

A revolution in farming had arrived.

Above: **John Deere Moline factory postcard**

Facing page: **John Deere Model D**
The first Model D tractors such as this 1923 "Spoker" used a 482-ci (7,895-cc) engine and a two-speed transmission. By 1926, displacement would increase to 501 ci (8,206 cc). A three-speed version debuted in 1934. Owners: Walter and Bruce Keller. (Photograph © Andy Kraushaar)

CHAPTER 1

Deere Enters the Tractor Field, 1914–1923

First Steps into the Tractor Business

Henry Ford's Fordson first took the farm by storm in 1918. Suddenly, a lightweight tractor was affordable and available for everyone. By the mid-1920s, 80 percent of the farm tractors at work around the globe were Fordsons.

Up until this time, Deere & Company had steadfastly remained an implement-making firm, selling other manufacturers' tractors—such as the Gas Traction Company's famed Big Four 30—alongside its own wares. But the visionaries at Deere's helm led by President William Butterworth foresaw that if the company was to survive and prosper, it needed a tractor of its own.

Deere's engineers experimented with various inventions, from competent but too-modern motor cultivators to three-wheeled contraptions that would have made Rube Goldberg proud. Finally, after many failures and few successes, Deere decided to buy tractor-building experience, acquiring all rights to the established and respected Waterloo Boy machine on March 14, 1918.

Then, in 1923, Deere launched its updated Model D version of the Waterloo Boy. The green machine was as hardy as a mule, as strong as an ox, and as trustworthy as a farm dog. Deere had entered the tractor business with a flourish.

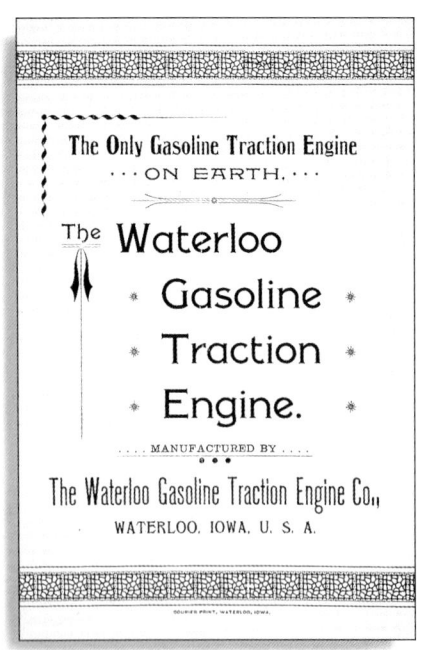

The Only Gasoline Traction Engine
· · · ON EARTH. · · ·

The Waterloo Gasoline Traction Engine.

. . . . MANUFACTURED BY . . . .

The Waterloo Gasoline Traction Engine Co.,
WATERLOO, IOWA, U. S. A.

Froelich Gas Tractor brochure

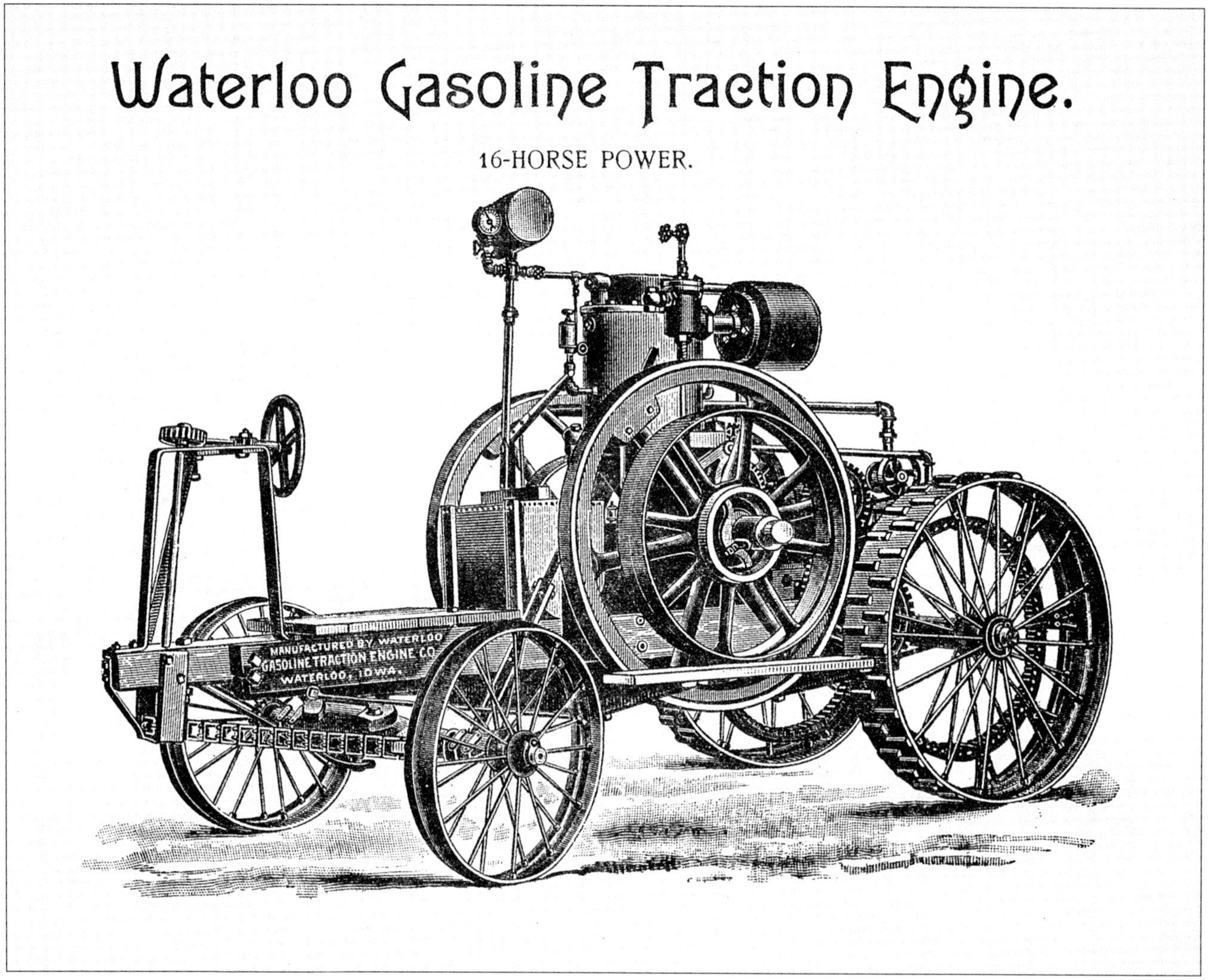

Froelich Gas Tractor

Iowa thresherman and inventor John Froelich created his gasoline traction engine in 1892 using parts from a steam tractor and a Van Duzen gasoline engine. The Froelich was the first gas tractor capable of powering itself forward and backward. This is an engraving of the machine from an original Waterloo Gasoline Traction Engine Company brochure.

John Deere lineup

Deere soon realized that tractors were the way of the future in farming and rushed to build or buy one of its own. In the early years, Deere dealers sold tractors from other companies—such as Gas Traction's Big Four 30 or Minneapolis's Universal Farm Motor, as shown here—to complete its lineup of agricultural products. This was dealer R. M. Addison & Son's exhibit at the Lyon County Fair, Minnesota, in 1910.

John Deere Motor Cultivator

Deere engineers crafted several prototype tractors in the mid 1910s but most of them were Rube Goldberg–type creations and were never produced. This two-row Motor Cultivator was invented by Walter Silver in 1916–1917 and designed to replace the horse on small farms.

Above: **John Deere–Dain All-Wheel-Drive**
Joseph Dain Sr. joined Deere's board of directors after Deere purchased his hay machinery firm in 1910. Dain was soon put in charge of developing a Deere tractor that could sell for $700. His All-Wheel-Drive of 1917 could be considered Deere's first tractor, although only one hundred were built. (Photograph © Andrew Morland)

Above right: **John Deere–Dain All-Wheel-Drive brochure**

Above: **Waterloo Boy advertisement**

Facing page: **Waterloo Boy Model N**
The Waterloo Gasoline Engine Company's Waterloo Boy tractor was advertised as the "one-man tractor" and was a true standout for its fine engineering. In fact, the tractor was so good, Deere bought out the firm in 1918 for $2.1 million and added the Waterloo Boy to its own line of farm machinery. Finally, Deere had a tractor of its own. (Photograph © Andrew Morland)

Both photos: **Waterloo Boy Model N**
Deere produced three succeeding styles of Waterloo Boys: The Model N in Styles A, B, and C from 1918 through 1924. Power came from a 465-ci (7,617-cc) two-cylinder engine creating 25 belt hp. This simple twin-cylinder engine would become the godfather of Deere's venerable Johnny Popper line, lasting in production through 1959. (Photograph by Hans Halberstadt)

Facing page: **John Deere Model D**
Deere engineers began updating the Waterloo Boy Model N Style C in the early 1920s. When the new version was ready, Deere decided to list the tractor simply as the John Deere Model D. This was one of the first, a 1923 "Spoker." Owners: Walter and Bruce Keller. (Photograph © Andy Kraushaar)

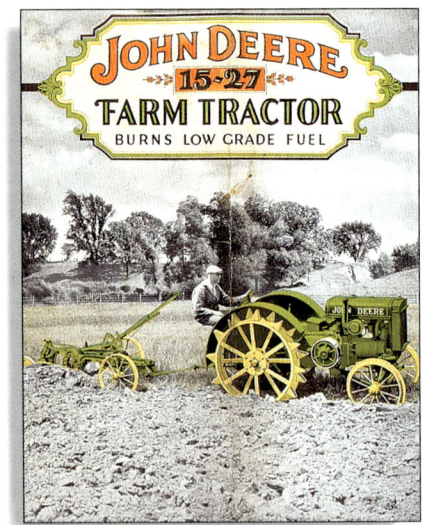

John Deere Model D brochure

John Deere Model DI
Deere's Model D Industrial version arrived in 1935 but only 100 were built during six years of production. The DI featured extended control levers, dual PTO outlets, sideways seating for viewing towed equipment, safety yellow paint, and optional final-drive gearing for low- and high-speed use. Owner: John Caes. (Photograph © Andy Kraushaar)

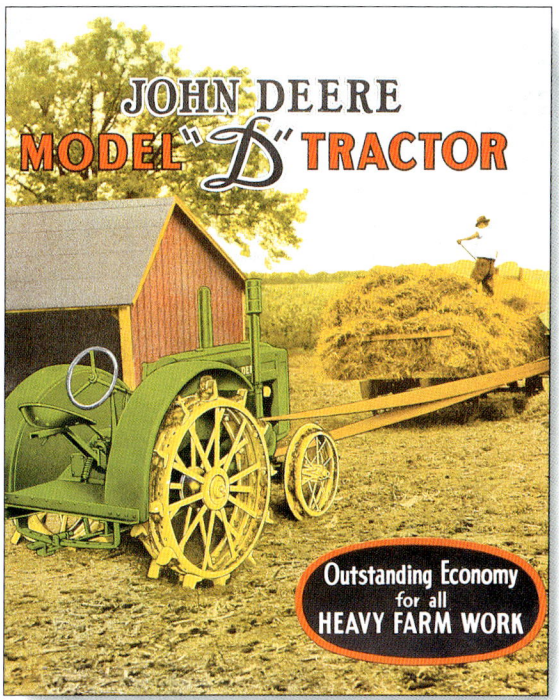

Above: **John Deere Model D brochure**

Left: **John Deere Model D**
The controls of the Model D were simplicity itself, and many a farm boy and girl learned to drive tractor from a seat such as this. (Photograph by Hans Halberstadt)

CHAPTER 2

Poppin' Johnnies, 1923–1937

The Beloved Sound of Two Cylinders

Deere inherited its two-cylinder technology from the Waterloo Gasoline Tractor Company and its Waterloo Boy machine. Two cylinders were simple and basic, inexpensive to operate and maintain, and powerful enough for most 1920s farms.

With the debut of International Harvester's Farmall in 1924, Deere rushed to expand its tractor line, adding its own general-purpose row-crop machine, known initially as the Model C and later as the GP. The Model GP soon evolved into a full lineup of machines, including the Models A and B, and later the G, H, and M.

By the mid 1920s, Deere's green machines were inexpensive enough that most every farmer could afford one and refined enough that it didn't take a wizard with a wrench and a large vocabulary of curses to operate it. "Power farming" was here to stay.

Along the way, the engine noise of the two-cylinder Deere also worked its way into farmers' hearts. It was a simple sound that you could trust and believe in, and Deeres became lovingly known as Poppin' Johnnies and Johnny Poppers.

Above: **John Deere's "The Handy Farm Account Book"**

Facing page: **John Deere Model BW**
Deere's Model B was unveiled in 1934 and became a landmark tractor for the firm. This wide-front Model BW dates from 1937. Owner: Robert Dufel. (Photograph © Andy Kraushaar)

Above: **John Deere Model C**
After International Harvester unleashed its revolutionary Farmall in 1924, Deere rushed to build its own general-purpose row-crop tractor. The result was the Model C, unveiled in 1926. Unfortunately, after just ninety-nine Model C tractors were built in 1927, problems meant they had to be recalled. When the tractor was re-released in 1928, it was titled the Model GP to avoid confusion on the telephone between the Models D and C. Owners: Walter and Bruce Keller. (Photograph © Andy Kraushaar)

Left: **John Deere Model GP**
Arriving in 1928, the Model GP was a refined version of the Model C. It still boasted the Model C's best points, however, including its radical four power outlets—PTO, drawbar, belt, and mechanical lift—making it an ideal two-to-four-row tractor. (Photograph © Andy Kraushaar)

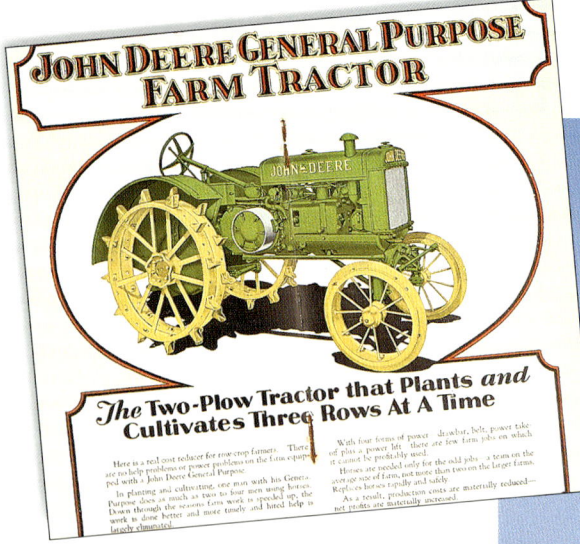

John Deere Model GP brochure

John Deere Model GP

The debut of the Model C/GP signaled a revolution at Deere & Company. Unveiled in 1926 as the C and revised in 1928 as the GP, this was Deere's first general-purpose row-crop tractor. Owner: Don Wolf. (Photograph © Andrew Morland)

Both photos: **John Deere Model GPO**
An Orchard version of the GP arrived in 1931 with the usual extra cowling to protect trees and their fruit from the tractor. About 700 were built. Some twenty-five copies of a crawler version known as the GPO-L were built by Lindeman Manufacturing in Yakima, Washington, in 1933–1934. (Photograph by Hans Halberstadt)

Above: **John Deere Model GPWT brochure**

Right: **John Deere Model GPWT**
A tricycle version of the GP was offered starting in 1928, and then followed by the Model GPWT with wide rear tread. This 1931 version featured side steering; it was superseded in 1932 by an overhead-steer version. Owner: Paul Steffes. (Photograph © Andy Kraushaar)

Facing page: **John Deere Model A group**
The Model A was a development of the GPWT, with production starting in March 1934. As historian Wayne G. Broehl Jr. wrote in *John Deere's Company*, "Probably no single stage in the entire history of the company's product development was any more important than this one [the launch of the Models A and B]." These Model A tractors date from 1934 through 1938. Owners: Howard and Bonnie Miller. (Photograph © Andy Kraushaar)

Above: **John Deere Models A and B brochure**

Left: **John Deere Model AWH**
The A was offered in a wide range of versions for various farm operations, including the narrow AN and wide AW, and Hi-Crop variations of both stances. This 1937 AWH stood atop 40-inch (100-cm) rear wheels instead of the standard 36-inch (90-cm). Owners: Walter, Bruce, and Jason Keller. (Photograph © Andy Kraushaar)

John Deere Model AI

The Industrial version of the A rode on a 7-inch (175-mm) shorter wheelbase than the standard AR and also included a heavy-duty drawbar, beefed-up rear axle, and cushioned seat. It was built in small numbers from 1936 through 1941; this is a 1939 machine. Owners: Walter, Bruce, and Jason Keller. (Photograph © Andy Kraushaar)

Above: **John Deere Standard Models brochure**

Left: **John Deere Model AR**
The standard-tread, or "Regular," Model AR arrived in 1935. It was originally known as the AS, for A Standard. Owners: Walter, Bruce, and Jason Keller. (Photograph © Andy Kraushaar)

Above: **John Deere Model AO**

The Orchard version of the A made its debuted in 1935 with bare-bones cowling protection. Only 861 of the first AO variation were built before the revised AOS arrived in 1936. Owner: Irv Baker. (Photograph by Hans Halberstadt)

Above: **John Deere Model AOS**

The AOS, or streamlined AO, arrived in 1936 riding atop the AI's shorter wheelbase and featured dramatically more-protective bodywork cowling then early AO. This is a 1937 machine. Owners: Walter, Bruce, and Jason Keller. (Photograph © Andy Kraushaar)

Facing page: **John Deere Model AO**

In 1940, the AO returned with a larger engine, now displacing 321 ci (5,258 cc) instead of the old 309 ci (5,061 cc). Power was also up, from 25 hp to 29.59 hp. Owner: Augie Scoto. (Photograph by Hans Halberstadt)

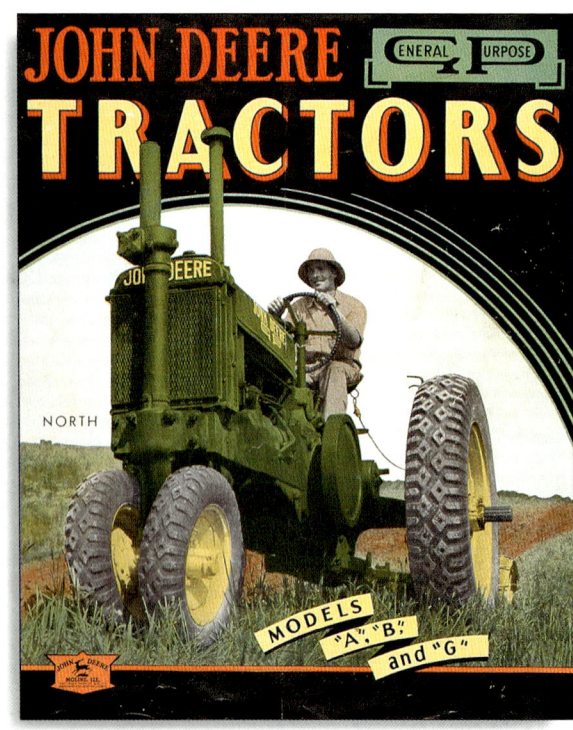

Above: **John Deere Models A, B, and G brochure**

Right: **John Deere Model B**
Soon after the Model B arrived in 1934, Deere's two-cylinder tractors became known variously as "Poppin' Johnnies" or "Johnny Poppers" as the engine's sound was so distinctive that farm wives could tell from the note of the engine when their husbands had idled down the tractor to come in for supper. This is a 1936 machine. (Photograph © Andrew Morland)

John Deere Model BWH-40
The frame of the early 1934–1937 Model B tractors was lengthened 5 inches (125 mm) in August 1937 to allow interchangeability of Model A mid-mounted equipment. This 1938 BWH-40 was a rare style using parts from the BO to create a narrow-rear-axle Hi-Crop variation. Owners: Walter and Bruce Keller. (Photograph © Andy Kraushaar)

John Deere Model BW
Operating the controls of a Model B became second nature to many farmers. (Photograph by Hans Halberstadt)

Above: **John Deere Model BI**
As with the Models D and A, Deere also released an Industrial variation of the B, in 1935. This is a 1937 machine. In addition to the wheeled models, Lindeman converted one BI to ride on treads. Owners: Walter and Bruce Keller. (Photograph © Andy Kraushaar)

Facing page: **John Deere Model BN**
Deere's Model BN with its narrow front end was known as the B "Garden Tractor." Like all 1934–1938 Model Bs, it was powered by a 149-ci (2,441-cc) engine producing 16 hp. This 1935 machine is fitted with aftermarket solid wheels. Owners: Bob and Mary Pollock. (Photograph © Andy Kraushaar)

Both photos this page: **John Deere Model BO**
The Orchard version of the Model B was based on the standard-tread BR. To protect grove trees from the tractor, guards and cowling covered the machine's filler caps, air intake, and rear wheels. (Photograph by Hans Halberstadt)

Facing page: **John Deere Model BO-L**
Lindeman modified about 1,700 Model BO tractors into BO-L crawlers starting in 1939. Production continued until 1946, when Deere bought out the Yakima, Washington, firm. (Photograph by Hans Halberstadt)

Above: **John Deere Model BO-L flyer**

Left: **John Deere Model G**
The Model A's larger sibling, the G, made its debut in late 1937 for the 1938 season. Power came from a 412.5-ci (6,757-cc) engine creating 35.9 hp via a four-speed transmission. The larger row-crop machine was only available with twin front wheels. This is a 1942 model. Owners: Del and Don Endres. (Photograph © Andy Kraushaar)

CHAPTER 3

Tractor Style Down on the Farm, 1938–1955

Streamlining and Ergonomics Refine the Mechanical Mule

Above: **John Deere on show**
The new styled Deere won approving gazes from everyone.

Facing page: **John Deere Model BW**
Styling came to the farmyard in 1938 with industrial designer Henry Dreyfuss's reworking of the early "unstyled" Poppin' Johnnies. This 1949 wide-front BW featured the new streamlined hood and front-end bodywork that was the hallmark of the "styled" Deeres. Owner: Seno Bast. (Photograph © Andy Kraushaar)

The first Johnny Poppers were simple and crude mechanical mules, and their styling was pure practicality. But in the 1930s, styling began to sell products, from fanciful automobiles to curvaceous Coca-Cola bottles. Not to be outdone, Deere decided to update its tractors with a facelift.

To handle the styling of their machines, Deere sought out the New York–based industrial-design firm of Henry Dreyfuss Associates in August 1937. Dreyfuss had never heard of Deere & Company, but he soon began drafting sketches redesigning the looks of the firm's tractors.

In 1938, the "styled" Models A and B made their debuts encased in streamlined bodywork that suddenly made them appear new and modern.

Dreyfuss's work didn't end there, however. The new Deeres were also more efficient, comfortable, and safer. Form and function were joined as never before on something as simple and crude as a farm tractor.

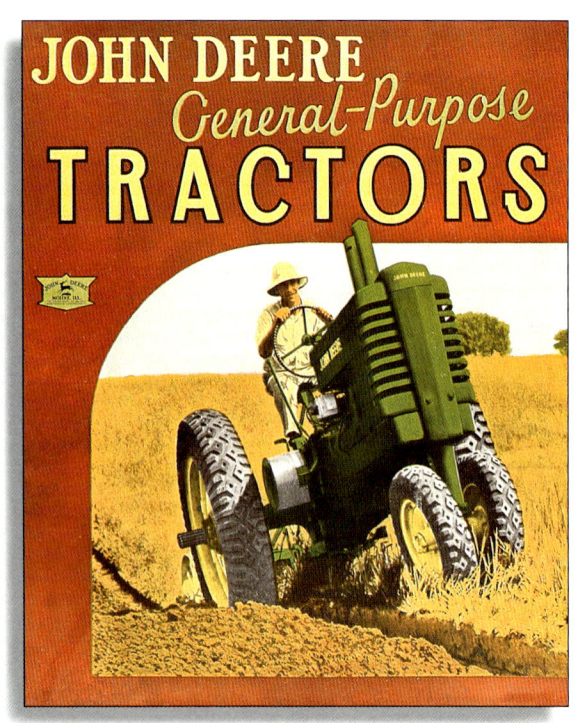

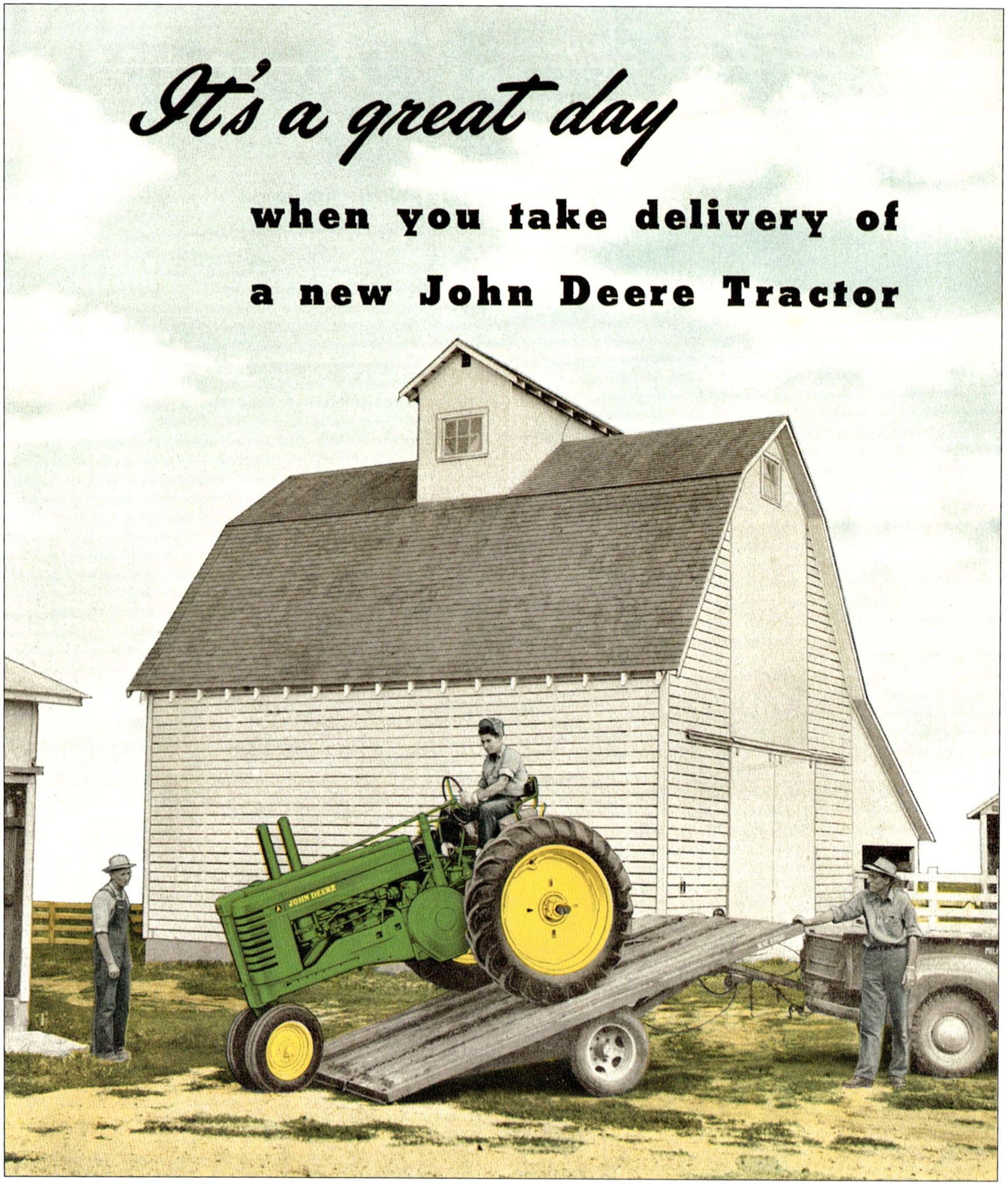

Above: **John Deere Models A and B brochure**

Right: **John Deere Model A brochure**

Facing page: **John Deere Model B**
The row-crop Models A and B were the first to receive Henry Dreyfuss's styling facelift. This 1939 B was also powered by a larger engine, giving the model a needed boost. The first styled Bs boasted 175-ci (2,867-cc) engines, but this was increased to 190 ci (3,112 cc) at serial number 201000. Owners: Del and Don Endres. (Photograph © Andy Kraushaar)

John Deere Model D
The Model D was the third Deere tractor to gain Dreyfuss's styling, also in 1939 following the Model A and B. This 1948 machine still retained the 501-ci (8,206-cc) engine of the earliest D. Owner: Bob Rettig. (Photograph © Andy Kraushaar)

John Deere Models G, GN, and GW brochure

John Deere Model G
The Model G was styled in 1942 and relaunched as the GM with a new six-speed transmission. The machine was renamed the G in 1947 and joined by the narrow GN, GH Hi-Crop, and wide GW variations. This is a 1951 machine. (Photograph © Andrew Morland)

Right: **John Deere Model A**
The styled Model A was given an increase in displacement in 1940 with a 321-ci (5,258-cc) engine now producing 29.59 hp. Early styled As had four-speed gearboxes, but a six-speed was soon available. This is a 1944 machine. Owners: Howard and Bonnie Miller. (Photograph © Andy Kraushaar)

Facing page: **John Deere Model As**
Pneumatic rubber tires had established their virtues by 1940. Yet during the World War II years, many tractors were equipped with the old-fashioned steel wheels as rubber was deemed a restricted wartime material. Owner: Jim Joas. (Photograph © Andy Kraushaar)

Above: **John Deere Model AH**
The ANH and AWH were left out of the new styled lineup in 1947, but a Hi-Crop styled Model A was launched in 1950, sharing many components with the high-clearance GH. This is a 1952 machine. Owner: Lloyd Sheffler. (Photograph © Andy Kraushaar)

Right: **"Tales of the Day"**
Artist Dave Barnhouse remembers the good old days of the Johnny Popper in this painting. (Original art by Dave Barnhouse © 2002 Hadley Licensing, Bloomington, MN)

Tom Brent and his Tractor
Many a 4-H youth learned the lessons of maintaining the family's Johnny Popper from this famous book that explained everything from engineering theory to hands-on repair.

John Deere Model L
The Model L arrived in 1937 to replace the earlier, one-year-only Models Y and 62. In 1938, the L also received Dreyfuss's styling but with a unique look that separated it from the row-crop machines. The L also differed from the others by its vertical rather than horizontal twin-cylinder engine. Displacement was 66 ci (1,081 cc). Owners: Bob and Mary Pollock. (Photograph © Andy Kraushaar)

Below: **John Deere Model LA**
The more-powerful LA arrived in 1940 with 14 hp rather than the L's 10.42 hp. The LA was also more versatile with its adjustable front axle, rear PTO, and electric lighting and starter. (Photograph by Hans Halberstadt)

Above: **John Deere Model LI**
The Industrial LI was commonly painted in high-visibility yellow paint rather than the standard Deere green. This is a 1945 machine. Owner: Herb Walter. (Photograph © Andy Kraushaar)

Left: **John Deere calendar image**

Facing page: **John Deere Model HNH**
Deere launched its small row-crop tractor, the Model H, in 1937 dressed in Dreyfuss's styling from day one. Power came from a 90.7-ci (1,486-cc) engine producing 14.84 hp via a three-speed gearbox. As with the early Model B, HN, HW, HNH, and HWH variations were soon available. This is a 1941 machine. Owners: Walter and Bruce Keller. (Photograph © Andy Kraushaar)

Above: **John Deere Model MI**
The Model M bade goodbye to the Model L and LA lines upon its debut in 1947. Power came from a vertical 100.5-ci (1,646-cc) engine with 20.45 hp and a four-speed transmission. This Industrial MI was painted in high-visibility yellow. Owners: Bob and Mary Pollock. (Photograph © Andy Kraushaar)

Facing page: **John Deere Model HWH**
Both the HNH and HWH rode atop the Model B's 38-inch (95-cm) rear wheels. This wide-front Model H Hi-Crop also boasted full hydraulics. Owners: Walter and Bruce Keller. (Photograph © Andy Kraushaar)

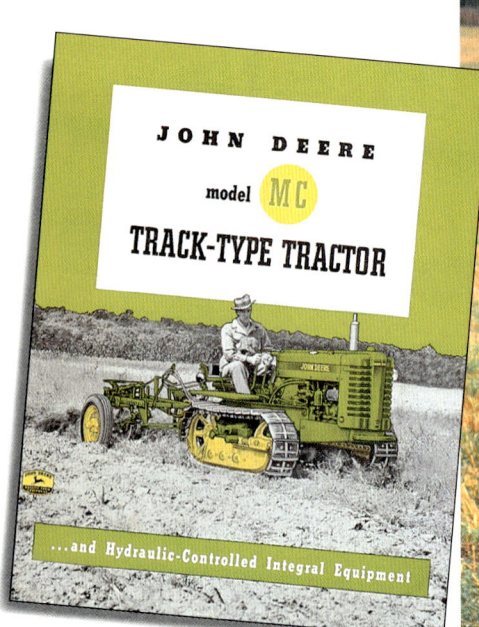

John Deere Model MC brochure

John Deere Model M
Along with the M and MI variations, the model was also available in the twin-front-wheel MT, single-front-wheel MTN, wide-front MTW, and MC crawler. This is a 1948 machine. Owner: Ken Koberg. (Photograph © Andy Kraushaar)

John Deere Model R brochure

John Deere Model R
Deere was a pioneer in experimenting with diesel power as far back as the 1930s. Yet the firm's first diesel machine was not released until 1949 when the Model R was launched. The two-cylinder overhead-valve diesel engine displaced 416 ci (6,814 cc) to create 50.96 hp. Originally designed as a replacement for the Model D, that venerable tractor was still so popular that it soldiered on alongside the R until 1953. (Photograph by Hans Halberstadt)

CHAPTER 4

Lanz, the German Cousin, 1911–1970

Joining Forces Across the Ocean

Heinrich Lanz was born in Germany in 1838, just one year after John Deere crafted his first steel plow. Like Deere, Lanz saw a great future in farming, and he began importing steam engines and threshing machines. In 1859, Lanz established his own factory in Mannheim and continued to add to his line of agricultural implements.

In 1902, Lanz visited the Deere works in Moline and met Charles Deere, the son of John Deere. This meeting sparked Lanz's enthusiasm for growth and paved the way for future cooperation between the two firms.

In 1921, Lanz's son, Karl, unveiled a stunning new tractor, the Bulldog, engineered by Dr. Fritz Huber as the world's first hot-bulb-fired tractor burning inexpensive crude oil. The Bulldog became a huge success for Lanz, with versions built throughout the world under license for decades thereafter. The 200,000th Bulldog rolled off the lines in 1956.

Also in 1956, Deere bought out the German firm and continued to build a long list of tractors at the famed Mannheim works through to the present. The Mannheim-made Deeres continue to be sold around the globe as well as being imported into the United States.

Above: **Lanz Bulldog brochure**

Facing page: **John Deere–Lanz 2816**
The Bulldog soldiered on under Deere ownership but now wore Deere's trademark green rather than Lanz's medium-blue paint scheme. This 1959 full-diesel 2816 was one of twelve Bulldogs available that year. Built from 1955 through 1960, it produced 28 hp.
(Photograph © Andrew Morland)

Above: **Lanz Bulldog HL brochure**

Right: **Lanz Bulldog HR9 EIL**
Introduced in 1938, the EIL Bulldog was equipped with road gears and, like Minneapolis-Moline's infamous UDLX Comfortractor, was designed to be driven to town after a day plowing fields. Powered by a massive 629-ci (10,303-cc) engine, the EIL had five or six gears for a top speed of 25 miles per hour (40 km/h). This is a 1939 machine. Owner: Pierre Bouillé. (Photograph © Andrew Morland)

Above: **Lanz Bulldog HR4 and HR5 brochure**

Left: **Lanz Bulldog D9506**
In 1950, the Bulldog line was restyled, giving it a more modern visage. The -06 suffix denoted six-speed models whereas the large-displacement -16 version had nine speeds. This is a 1952 machine. (Photograph © Andrew Morland

Lanz Bulldog D2806 brochure

Above: **Lanz Bulldog D2806**
This six-speed 1954 D2806 was powered by a 226-ci (3,702-cc) single-cylinder engine producing 28 hp. (Photograph © Andrew Morland)

Facing page: **Lanz Bulldog HR9 EIL**
Lanz's hot-bulb-fired, crude-oil-burning Bulldog sparked a revolution in European agriculture upon its debut in 1921. Licensed versions of the machine were built in Italy, Spain, France, Czechoslovakia, Great Britain, and elsewhere around the globe. (Photograph © Andrew Morland)

Lanz Bulldog D5016 brochure

Above: **Lanz Bulldog D6016**
The 1955–1962 D6016 was at the top of Lanz's line. Its one-cylinder full-diesel engine pumped out a muscular 60 hp via a nine-speed gearbox. This is a 1959 machine. (Photograph © Andrew Morland)

Right: **John Deere–Lanz 300**
Deere's New Generation of Power arrived on European farms in January 1960, six months before it was launched in the United States. The Mannheim-built 300 and 500 boasted four-cylinder diesel engines producing 30.8 hp (28 PS) and 39.5 hp (36 PS) respectively. (Photograph by Chester Peterson Jr.)

CHAPTER 5

The Numbered Series Years, 1952–1960

A Good Thing Gets Better

The years following World War II were the best of times and the worst of times for the farm tractor. Tractors were now viewed as an essential part of the farm, and more tractors were being built than ever before. Mechanical refinements such as power takeoffs, hydraulic systems, diesel engines, and ergonomics continued to make better machines.

Deere designers and engineers continued to improve the two-cylinder line with upgrades and refinements over the decade. The Letter Series gave way to the Number Series, followed by the 20 Series, and then the ultimate two-cylinder Deeres, the 30 Series.

Deere's two-cylinder tractors had enjoyed a long life, but at the same time, the company began looking to the future in creating a new generation of tractors. Many Deere stalwarts from dealers to farmers certainly had a hint of changes in the wind, but even with the debut of the radical 8010 in 1959, few folk guessed how momentous they would be.

Above: **John Deere Model 70 Diesel brochure**

Facing page: **John Deere Model 530**
The venerable Letter Series was updated as the Number Series in 1952. The change in name primarily reflected subtle mechanical refinements rather than revolutions, and beyond the new numbering on the radiator cowling, the tractors largely continued the same. (Photograph © Andy Kraushaar)

John Deere 20 Series brochure

Above: **John Deere Model 40 Crawler brochure**

Facing page: **John Deere Model 50**
The Model B was upgraded as the 50 in 1952. Like the replacement for the Model A, the 60, it now featured a cast frame rather the former pressed steel. By 1955, LPG fuel was an option. (Photograph © Andy Kraushaar)

They're the only truly modern tractor cultivators in the field

BRING YOUR cultivating up to date with a John Deere Two- or Four-Row *Quik-Tatch* Cultivator. Modern design places the John Deere head and shoulders above any other cultivator in the field. Interpreted into performance, a John Deere does better work at lower cost in all row crops and field conditions. Faster, easier attaching and detaching without disturbing row spacings . . . quicker, more positive dodge . . . greater clearance . . . better visibility . . . greater comfort for the operator . . . easier handling with effortless hydraulic control . . . fewer, simpler adjustments . . . greater strength . . . longer-wearing, better-scouring sweeps and shovels . . . and a complete line of regular and special equipment for better work and for adapting your John Deere Cultivator to other jobs—all these ten valuable performance features are found *only* in John Deere *Quik-Tatch* Cultivators.

Performance That Means Money

The outstanding performance of John Deere Two- and Four-Row Cultivators means more dollars in your pocket in time and effort saved as well as greater harvest profits from cleaner fields. The "Quik-Tatch" feature alone brings big savings in time and, by making the tractor more readily and easily available, permits broader use of your tractor during one of the busiest seasons of the year.

The clean design of John Deere Tractors plays an important part in the extra visibility, clearance, comfort, and ease of handling in a John Deere *Quik-Tatch* Cultivator. Tractor and cultivator make up an unbeatable team with more years of service and lower upkeep costs.

Let's examine the many outstanding advantages of John Deere *Quik-Tatch* Cultivators on the following pages. You'll soon appreciate why no other cultivators "stack up" with them by any measure of value.

JOHN DEERE

2- and 4-ROW QUIK-TATCH CULTIVATORS

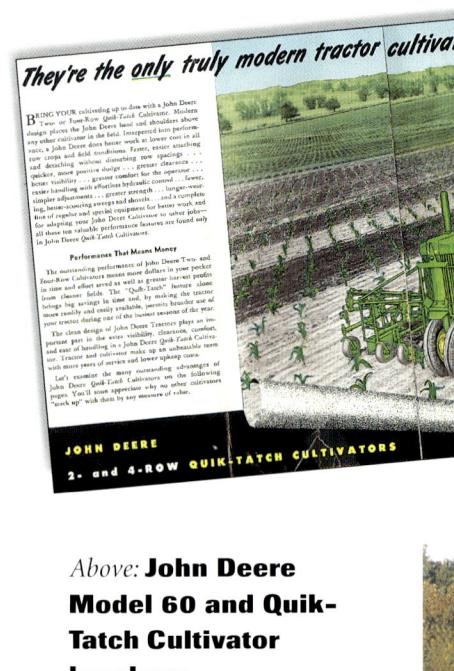

Above: **John Deere Model 60 and Quik-Tatch Cultivator brochure**

Right: **John Deere Model 60**

The 60 replaced the A and was available in Orchard, Hi-Crop, and the newly designated S, or Standard, model. 38 hp was now on tap. (Photograph © Andy Kraushaar)

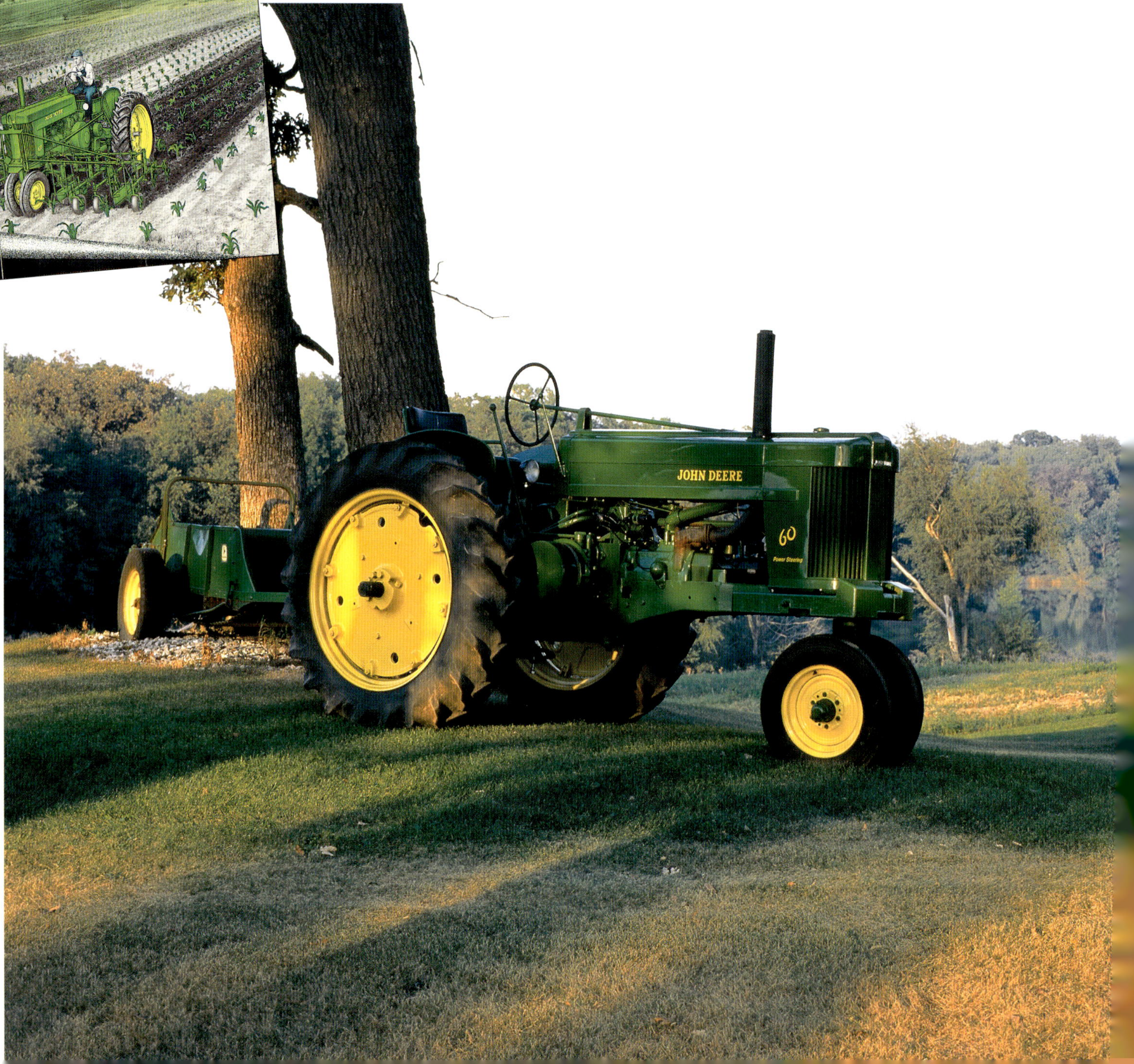

Both photos: **John Deere Model 60 Orchard**
The Model 60O was launched in 1954, two years after the row-crop version. Protective cowling was minimal. (Photograph by Hans Halberstadt)

John Deere Models 50, 60, and 70 brochure

Above: **John Deere Model 70**
The Model 70 replaced the G in 1953, offering a staunch 50.35 belt hp and cutting into Model 60 sales. In 1954, the 70D became Deere's first diesel row-crop machine, with 45.7 drawbar hp. This is a 1956 machine. (Photograph © Andrew Morland)

Facing page: **John Deere Model 80**
The Model R was superseded by the 80, which also got a power boost to lift it above the new 70D. Displacement was now 472 ci (7,731 cc), producing 67.6 PTO hp. (Photograph © Andy Kraushaar)

Above: **John Deere Models 520, 620, and 720 brochure**

Right: **John Deere Model 520**
Deere engineers revised the firm's tractor lineup in 1956 with a 20 percent power increase to keep up with demands from farmers and their ever-growing farm acreages. This 20 percent increase was duly reflected in the new tractors' names, the 20 Series. Thus, the 520 was simply a Model 50 with 20.45 PTO hp from its 100.5-ci (1,646-cc) engine. (Photograph © Andrew Morland)

Above: **John Deere Model 620 Grove brochure**

Left: **John Deere Model 620**

The 60 was recast as the 620, featuring a shorter stroke but increased engine speed. Displacement was now 302.9 ci (4,962 cc), and power was up to 35.68 belt hp. (Photograph © Andy Kraushaar)

Above: **John Deere Model 820 brochure**

Left: **John Deere Model 820**
Deere's 820 diesel now produced 75.6 belt hp from its 472-ci (7,731-cc) engine. (Photograph © Andy Kraushaar)

Facing page: **John Deere Model 720**
The massive 720 was Deere's largest row-crop machine, boasting a 376-ci (6,159-cc) engine and 56.84 belt hp. Again, a diesel version was offered. (Photograph © Andy Kraushaar)

Above: **John Deere Models 530, 630, and 730 brochure**

Right: **John Deere Model 530**
A minor styling update heralded the debut of the 30 Series just two to three years after the 20 Series's arrival. Flat-topped fenders, a crisp new yellow-and-green paint scheme, and other minor changes marked this series as new for 1959. (Photograph © Andrew Morland)

John Deere Model 830
The muscular 830 replaced the 820. It was soon joined by an 830 Industrial version, 830 Super, and the 840. (Photograph by Chester Peterson Jr.)

CHAPTER 6

Big Green and the New Generations, 1959–1970

The Dawn of a New Era at Deere

The New Generation of Deere tractors made its dramatic debut in Dallas, Texas, in August 1960. The venerable two-cylinder Poppin' Johnny had been retired, and a full line of four- and six-cylinder tractors had arrived to replace it.

These new tractors were modern and magnificent machines that set the direction of the world's tractor market for the future. Ergonomics, advanced safety features, and continual development of mechanical upgrades marked these Deeres as the most efficient, hard-working tractors available.

With the New Generation, Deere made a key move. The line of tractors that not only changed the face of farming at the time but also put Deere in a position to survive the worst ravages of the rollercoaster agricultural economy and set the stage for Deere to become the world's largest tractor manufacturer.

Above: **John Deere Model 8010 brochure**

Facing page: **John Deere Model 8020**
The debut of the New Generation of Power was foreshadowed in 1959 by the launch of Deere's radical 8010 and its successor, the 8020. These four-wheel-drive articulated tractors with their powerful six-cylinder diesels were not a success at the time, but the style they set would soon dominate the market. (Photograph by Chester Peterson Jr.)

Above: **John Deere Models 3010 and 4010 brochure**

Above: **John Deere Models 3010 and 4010**
Deere's 3010 and 4010 tractors were landmark machines that would become some of the most copied tractor designs of the 1960s and 1970s. They brought advances in ergonomics, safety, and all-round efficiency to the farmyard, and farmers swore by the curvaceous new Deeres. (Photograph by Chester Peterson Jr.)

Facing page: **John Deere Model 8010**
The 8010 was powered by a massive GM six-cylinder two-cycle diesel of 425 ci (6,962 cc). Power was an incredible 215 hp via a nine-speed gearbox. At the time, few farmers needed such muscle, but Deere was simply too early with too much. By the 1970s, such power would be in high demand. (Photograph by Chester Peterson Jr.)

Above: **John Deere Model 4010**
This cutaway show Model 4010 provided a welcome anatomy lesson in the advances heralded by the tractor. The six-cylinder engine produced an economic 80 hp run on gasoline. A diesel version was soon also available. (Photograph by Chester Peterson Jr.)

Facing page: **"The Rematch"**
A Deere and a Farmall prepare to fight it out in a friendly tug of war match in this painting by artist Dave Barnhouse. (Original art by Dave Barnhouse © 2002 Hadley Licensing, Bloomington, MN)

Above: **John Deere Model 2010**

The 2010 was powered by a 145-ci (2,375-cc) four producing 45 hp on gas. An LPG option was also available. Power was transmitted to the field via an eight-speed Synchro-Range gearbox. (Photograph by Chester Peterson Jr.)

Above: **John Deere Model 1010**

The little 1010 replaced the 430 but now boasted four instead of just two cylinders. A handy 35 hp was on tap, and a diesel engine was optional. (Photograph by Chester Peterson Jr.)

Facing page: **John Deere Models 4020, 3020, and 2520**

More power arrived with the upgraded 4020, 3020, and 2520. The flagship 4020 now featured a 404-ci (6,618-cc) six with 94 hp and an eight-speed transmission. (Photograph by Chester Peterson Jr.)

Both photos: **John Deere Model 4020H**
The 4020 Hi-Crop offered height and power to farmers needing higher clearance over tall crops. Owner: Glen Knudson. (Photograph by Chester Peterson Jr.)

Left: **John Deere Model 3120 brochure**

Below: **John Deere Model 7020**
The vision of the future that Deere unveiled with its 8010 and 8020 returned in 1970 in the guise of the 7020. This row-crop monster featured a turbocharged and intercooled six-cylinder engine delivering 146 PTO hp. The shape of the future of farm tractors had arrived. (Photograph by Chester Peterson Jr.)

Bibliography

Baldwin, Nick and Andrew Morland. *Classic Tractors A to Z*. Stillwater, Minn.: Voyageur Press, 1998.

Broehl, Wayne G. Jr. *John Deere's Company: A History of Deere & Company and Its Times*. N.Y.: Doubleday & Company, Inc., 1984.

Brown, Theo. *Deere & Company's Early Tractor Development*. Grundy Center, Iowa: Two-Cylinder, 1997.

Dregni, Michael, ed. *This Old John Deere: A Treasury of Vintage Tractors and Family Farm Memories*. Stillwater, Minn.: Voyageur Press, 2002.

Macmillan, Don. *The Big Book of John Deere Tractors: The Complete Model-by-Model Encyclopedia . . . Plus Brochures and Collectibles*. Stillwater, Minn.: Voyageur Press, 1999.

Macmillan, Don. *The Field Guide to John Deere Tractors*. Stillwater, Minn.: Voyageur Press, 2002.

Miller, Merle L. *Designing the New Generation John Deere Tractors*. St. Joseph, Mich.: American Society of Agricultural Engineers, 1999.

Sanders, Ralph W. *Ultimate John Deere: The History of the Big Green Machines*. Stillwater, Minn.: Voyageur Press, 2001.

John Deere plow advertisement

John Deere Clubs, Magazines, and Resources

Clubs and Magazines

Green Magazine
Richard Hain
2652 Davey Road
Bee, NE 68314–9132

John Deere Tradition
1503 SW 42nd Street
Topeka, KS 66609

Two-Cylinder Magazine
P. O. Box 10
Grundy Center, IA 50638–0010

Farm Collector
1503 SW 42nd Street
Topeka, KS 66609

General Parts

Ag Tractor Supply
Box 276
Stuart, IA 50250

All Parts International, Inc. (API)
3215 West Main Avenue
Fargo, ND 58103
www.stpc.com

Bob Martin Antique Tractor Parts
5 Ogle Industrial Drive
Vevay, IN 47043
www.venus.net~martin

The Brillman Company
Box 333
Tatamy, PA 18085
www.brillman.com

Central Michigan Tractor & Parts
2713 N. U.S. 27
St. Johns, MI 48879

Central Plains Tractor Parts
712 North Main Avenue
Sioux Falls, SD 57102

Colfax Tractor Parts
Rt. 1, Box 119
Colfax, IA 50054

Dengler Tractor
6687 Shurz Road
Middletown, OH 45042

Detwilier Sales
S3266 Highway 13 S
Spenser, WI 54479
715-659-4252

Discount Tractor Supply
Box 265
Franklin Grove, IL 61031

Draper Tractor Parts, Inc.
Rt. 1, Box 41
Garvield, WA 99130

Fresno Tractor Parts
3444 West Whitesbridge Road
Fresno, CA 93706

Iowa Falls Tractor Parts
Rt. 3, Box 330A
Iowa Falls, IA 50126

Dennis Polk Equipment
72435 SR 15
New Paris, IN 46553
www.dennispolk.com

Restoration Supply Co.
Dept. AP96 Mendon Street
Hopedale, MA 01747

Shepard's 2 Cylinder Parts, Service & Repair
John Shepard
E633-1150 Avenue
Downing, WI 54734

South-Central Tractor Parts
Rt. 1, Box 1
Leland, MS 38756

Southeast Tractor Parts
Rt. 2, Box 565
Jerfferson, SC 29718

Steiner Tractor Parts, Inc.
G-10096 South Saginaw Road
Holly, MI 48442
www.steinertractor.com

Surplus Tractor Parts Corp.
Box 2125
Fargo, ND 58107

Yesterday's Tractors
P. O. Box 160
Chicacum, WA 98325
www.ytmag.com

Specialized Parts

A-1 Leather Cup and Gasket Company
2103 Brennan Avenue
Fort Worth, TX 76106

Agri-Services
13899 North Road
Alden, NY 14004
Specializing in wiring harnesses

Proud Pa with Junior already at the controls.

John Deere–Dain All-Wheel-Drive

Antique Gauges, Inc.
12287 Old Skipton Road
Cordova, MD 21625
Specializing in gauges

Burrey Carburetor Service
18028 Monroeville Road
Monroeville, IN 46773
Specializing in Deere carburetors and governors

Dave Geyer
1251 Rohret Road SW
Oxford, IA 52322
Specializing in Deere tractor hoods

Jorde's Decals
935 Ninth Avenue NE
Rochester, MN 55906
www.jordedecals.com
Decals for John Deere

John R. Lair
413 L.Q. P Avenue
Canby, MN 56220
Specializing in Deere fenders

Lubbock Gasket & Supply
402 19th Street, Dept. AP
Lubbock, TX 79401

Jack Maple
Rt. 1, Box 154
Rushville, IN 46173
Decals for a wide variety of applications and models

M. E. Miller Tire Co.
17386 State Highway 2
Wauseon, OH 43567
www.millertire.com

Nielsen Spoke Wheel Repair
Herb Nielsen
3921 230th Street
Esterville, IA 51334

Olson's Gaskets
3059 Opdal Road E
Port Orchard, WA 98366
www.olsonsgaskets.com

Omaha Avenue Radiator Service
100 East Omaha Avenue
Norfolk, NE 68701

Roy Ritter
15664 County Road 309
Savannah, MO 64485
Specializing in Deere two-cylinder pumps

2-Cylinder Diesel Shop
Roger and Dana Marlin
Rt 2, Box 241
Conway, MO 65632
Specializing in Deere diesel repairs

Tractor Steering Wheel Recovering and Repair
1400 121st Street W
Rosemount, MN 55068

Tractor Manuals

Clarence L. Goodburn Literature Sales
101 West Main
Madelia, MN 56062

Intertec Publishing
P. O. Box 12901
Overland Park, KS 66282
www.intertecbooks.com

Jensales Inc.
P. O. Box 277
Clarks Grove, MN 56016
www.jensales.com

King's Books
P. O. Box 86
Radnor, OH 43066

Yesterday's Tractors
P. O. Box 160
Chicacum, WA 98325
www.ytmag.com

Index

Big Four 30, *9, 11*
Broehl, Wayne G. Jr., *26*
Butterworth, William, *9*
Dain, Joseph Sr., *13*
Deere & Company, *9, 24, 43*
Deere tractor models,
    All-Wheel-Drive, *13, 94*
    Model 1010, *88*
    Model 2010, *88*
    Model 2520, *88*
    Model 3010, *85*
    Model 3020, *88*
    Model 3120, *91*
    Model 40 Crawler, *71*
    Model 4010, *85–86*
    Model 4020, *88*
    Model 4020H, *90*
    Model 50, *71, 74, 76*
    Model 520, *76*
    Model 530, *69, 80*
    Model 60, *71–74, 77*
    Model 60 Orchard, *72, 73, 74*
    Model 62, *52*
    Model 620, *76–77*
    Model 620 Grove, *77*
    Model 630, *80*
    Model 70, *74*
    Model 70 Diesel, *69, 74*
    Model 7020, *91*
    Model 720, *76, 79*
    Model 730, *80*
    Model 80, *74*
    Model 8010, *69, 83, 91*
    Model 8020, *83, 91*
    Model 820, *79, 81*
    Model 830, *81*
    Model 830 Industrial, *81*
    Model 830 Super, *81*
    Model 840, *81*
    Model A, *21, 26, 29–34, 35, 36, 41, 43–44, 46, 48–50, 71, 72*
    Model AH, *50*
    Model AI, *30, 32*
    Model AN, *29*
    Model ANH, *50*
    Model AO, *32*
    Model AOS, *32*
    Model AR, *30, 31*
    Model AS, *31*
    Model AW, *29*
    Model AWH, *29, 50*
    Model B, *21, 29, 34–41, 43–44, 46, 55, 57*
    Model BI, *36*
    Model BN, *36*
    Model BNH, *4*
    Model BO, *35, 38*
    Model BO-L, *38, 41*
    Model BR, *38*
    Model BW, *21, 36, 43*
    Model BWH-40, *35*
    Model C, *21, 23, 24*
    Model D, *9, 16–19, 23, 36, 46, 59*
    Model DI, *18*
    Model G, *21, 34, 41, 47, 74*
    Model GH, *47, 50*
    Model GM, *47*
    Model GN, *47*
    Model GP, *7, 21, 23–26*
    Model GPO, *25*
    Model GPO-L, *25*
    Model GPWT, *26*
    Model GW, *47*
    Model H, *21, 55*
    Model HN, *55*
    Model HNH, *55, 57*
    Model HW, *55*
    Model HWH, *55, 57*
    Model L, *52–53, 57*
    Model LA, *53, 57*
    Model LI, *53*
    Model M, *21, 57–58*
    Model MC, *58*
    Model MI, *57, 58*
    Model MT, *58*
    Model MTN, *58*
    Model MTW, *58*
    Model R, *59, 74*
    Model Y, *52*
    Motor Cultivator, *12*
Deere tractor series,
    20 Series, *69, 71, 76, 80*
    30 Series, *69, 80*
    Letter Series, *69*
    New Generation of Power, *66, 83*
    Number Series, *69*
Deere, Charles, *61*
Deere, John, *61*
Deere–Lanz tractor models,
    2816, *61*
    300, *66*
    500, *66*
Dreyfuss, Henry, *43, 44, 46, 52, 55*
Farmall, *21, 23*
Ford, Henry, *9*
Fordson, *9*
Froelich Gas Tractor, *10*
Froelich, John, *10*
Gas Traction Co., *9, 11*
Henry Dreyfuss Associates, *43*
Huber, Dr. Fritz, *61*
International Harvester, *21, 23*
Lanz tractor models,
    Bulldog D2806, *64*
    Bulldog D5016, *66*
    Bulldog D6016, *66*
    Bulldog D9506, *63*
    Bulldog HL, *62*
    Bulldog HR4, *63*
    Bulldog HR5, *63*
    Bulldog HR9 EIL, *62, 64*
Lanz, Heinrich, *61*
Lanz, Karl, *61*
Lindeman Mfg., *25, 36*
Minneapolis Co., *11*
Minneapolis-Moline, *62*
Silver, Walter, *12*
Universal Farm Motor, *11*
Waterloo Boy tractor models, *9, 21*
    Model N, *15, 16*
    Model N Style C, *16*
Waterloo Gasoline Traction Engine Co., *10, 15*
Waterloo Gasoline Tractor Co., *21*

Forerunner of the Modern Line
of John Deere Tractors

Few farmers will remember the Froelich tractor of 1892—one of the first farm tractors to be powered with an internal combustion engine and the first to propel itself backward as well as forward. Never produced in volume, John Froelich's tractor did provide the dormant seed that later sprouted and brought forth the "Waterloo Boy" line.

TODAY *Leaders in the Features You Want*

The Waterloo, Iowa, factory in which the "Waterloo Boy" was built, purchased by John Deere in 1918, saw the development of the first John Deere Tractor—the Model "D," a two-cylinder tractor, famous for its simplicity, durability, ease of operation, and low operating costs. Today, the greatly expanded line of modern John Deere Tractors includes seven power sizes and twenty models to meet the exacting requirements on farms everywhere. Today, in every agricultural section of our country, John Deere Tractors are known among farmers for their dependable and economical performance.